Never to old to play with bricks...

Martin Ludwig & Frank Müller

WW2 Wehrmacht custom building instructions

to be build out of LEGO® Bricks

www.tredition.de

© 2015 Martin Ludwig & Frank Müller
Umschlag, Illustration: Frank Müller

Verlag tradition GmbH, Hamburg

ISBN
Paperback: 978-3-7323-4183-2
Hardcover: 978-3-7323-4184-9
e-Book: 978-3-7323-4185-6

Printed in Germany

Das Werk, einschließlich seiner Teile, ist urheberrechtlich geschützt. Jede Verwertung ist ohne Zustimmung des Verlages und des Autors unzulässig. Dies gilt insbesondere für die elektronische oder sonstige Vervielfältigung, Übersetzung, Verbreitung und öffentliche Zugänglichmachung.

Keine original LEGO® Bauanleitungen.
LEGO® hat dieses Buch weder gesponsert noch autorisiert. Die Benutzung des LEGO® Warenzeichens erfolgt zur eindeutigen Identifikation der LEGO® Produkte und soll keine Verletzung der Schutzrechte darstellen.
Alle hier abgebildeten Modelle sind bei der DPMA registriert.

Ebenfalls im tredition Verlag erschienen:
Bauanleitung Tiger & Königstiger

Also published by tredition:
Building instruction Tiger & Kingtiger tank

ISBN
Paperback: 978-3-7323-1027-2
Hardcover: 978-3-7323-1028-9
e-Book: 978-3-7323-1029-6

Martin Ludwig und Frank Müller sind zwei begeisterte Hobby AFOLs (Adult Fan of LEGO®) die sich auf den Nachbau von Militär Fahrzeugen spezialisiert haben. Die Fahrzeuge sind passend für die Größe der Figuren ausgelegt.

Alles begann mit einer Idee. Nach erfolgreichem Verkauf einiger Fahrzeuge bei eBay® wurde die Produktpalette stetig erweitert. Je nach Größe dauert das Design eines Fahrzeuges einige Wochen. Hier achten wir zum größten Teil auf die Funktionalität und die Verfügbarkeit der Teile. Natürlich können wir alles noch besser und komplexer bauen, aber nicht jedem Kunden stehen unsere Möglichkeiten zur Verfügung. Nach der Veröffentlichung des ersten Baubuches liegt der Fokus dieser Bauanleitungen auf kleineren Fahrzeugen. In diesem Buch enthalten sind die Bauanleitungen für:

- 7,5cm Panzerabwehrgeschütz (Seite 7 - 17)
- 2cm Vierlings Flak (Seite 18 - 27)
- Motorrad mit Beiwagen (Seite 28 - 41)
- Kradmelder Motorrad (Seite 42 - 49)
- NSU Kettenkrad (Seite 50 - 68)

Martin Ludwig and Frank Müller are two hobby LEGO® builders from Germany. They spend their time building military models out of LEGO® bricks. It all started with an idea. After selling a few vehicles on eBay®, they started a successful business. The design of each vehicle takes a few weeks. Functionality and accurate size of vehicles are the main goals. After the successful release of their first book, they have now finished their second instruction book focused on smaller vehicles. This book features:

- 7.5cm Anti-Tank Gun (page 7 - 17)
- 2cm Anti-Aircraft Gun (page 18 - 27)
- Motorcycle with Sidecar (page 28 - 41)
- Single Seat Motorcycle (page 42 - 49)
- Chain Motorcycle (page 50 - 68)

7,5cm Panzerabwehrgeschütz

Zum Bau des Modells benötigen Sie ca. 70 LEGO® Bausteine.
Länge: ca. 17,3 cm Breite: ca. 6,5 cm Höhe: ca. 4,7 cm

7.5cm Anti-Tank Gun

Requires approx. 70 LEGO® bricks.
length: ca. 17.3 cm width: ca. 6.5 cm height: ca. 4.7 cm

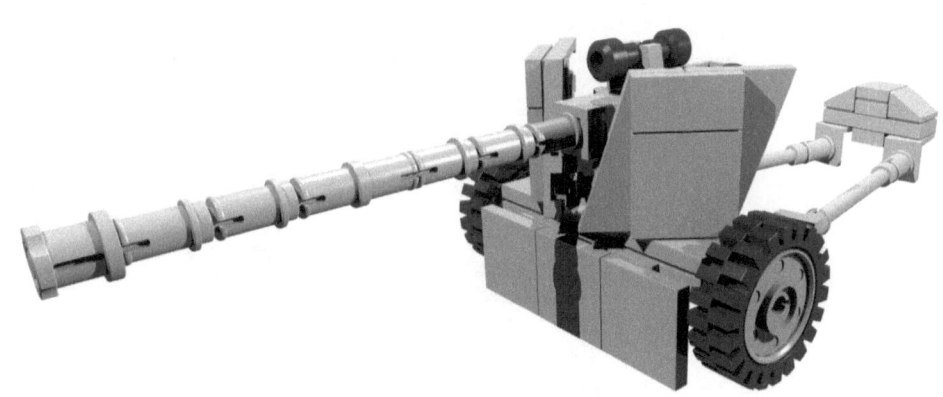

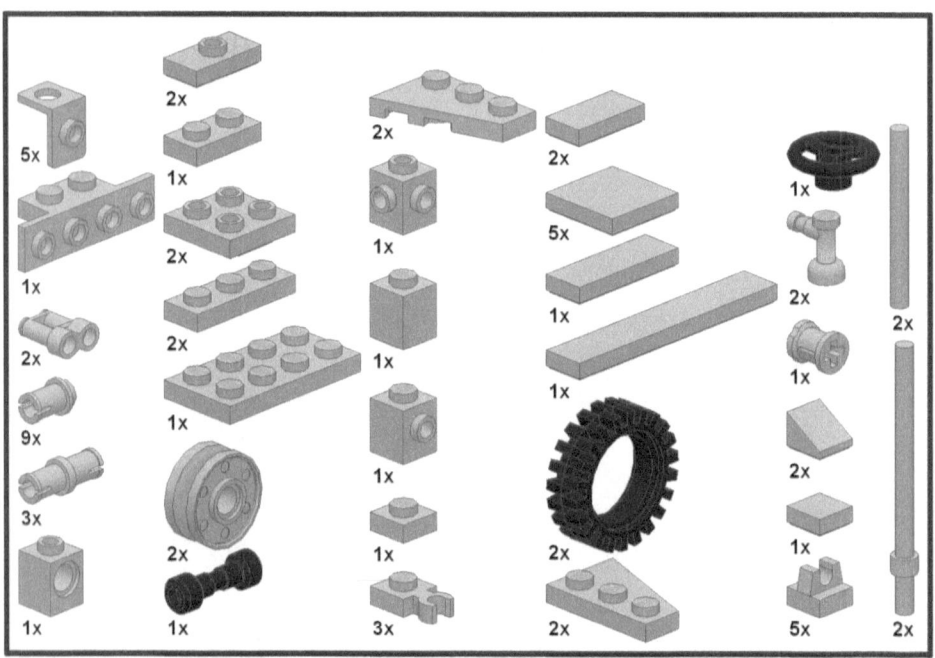

BLink ID	Color	Qty	Description
3020	Grey	1	Plate 2 x 4
3023	Grey	1	Plate 1 x 2
2436	Grey	1	Bracket 1 x 2 - 1 x 4 with Square Corners
2444	Grey	2	Plate 2 x 2 with Hole and Split Underside Ribs
3673	Grey	3	Technic Pin
3794b	Grey	2	Plate 1 x 2 with Groove with 1 Centre Stud
3068b	Grey	5	Tile 2 x 2 with Groove
6636	Grey	1	Tile 1 x 6
3623	Grey	2	Plate 1 x 3
42446	Grey	5	Bracket 1 x 1 - 1 x 1
4085d	Grey	3	Plate 1 x 1 with Clip Vertical (Thick C-Clip)
2555	Grey	5	Tile 1 x 1 with Clip
30162	Grey	2	Minifig Tool Binoculars Town
4274	Grey	9	Technic Pin 1/2
63965	Grey	2	Bar 6L with Thick Stop
3024	Grey	1	Plate 1 x 1
3070b	Grey	1	Tile 1 x 1 with Groove
63290	Grey	2	Slope Brick 31 1 x 1 x 0.667
3005	Grey	1	Brick 1 x 1
4733	Grey	1	Brick 1 x 1 with Studs on Four Sides
47905	Grey	1	Brick 1 x 1 with Studs on Two Opposite Sides
6541	Grey	1	Technic Brick 1 x 1 with Hole
30663	Black	1	Car Steering Wheel Large
45918	Black	1	Wheels Skateboard
63864	Grey	1	Tile 1 x 3 with Groove
4599	Grey	2	Tap 1 x 1 with Hole in Spout
30374	Grey	2	Bar 4L Light Sabre Blade
6590	Grey	1	Technic Bush with Two Flanges
56902	Grey	2	Wheel Rim 8 x 18 with Deep Center Groove
3483	Black	2	Tyre 7/ 56 x 17 Offset Tread
43722	Grey	2	Wing 2 x 3 Right
3069b	Grey	2	Tile 1 x 2 with Groove
43723	Grey	2	Wing 2 x 3 Left

www.ww2custombrickmodels.de

1

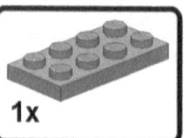

2

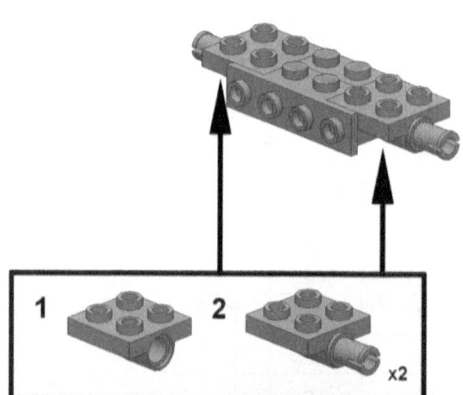

3

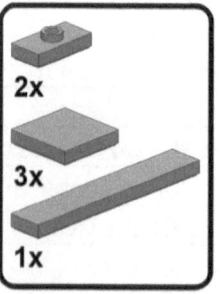

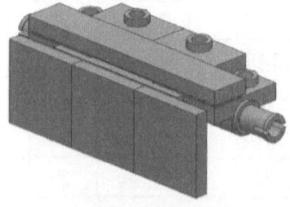

www.ww2custombrickmodels.de

4

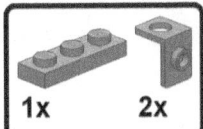

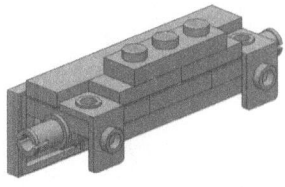

5

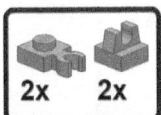

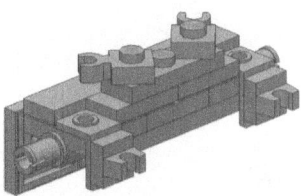

6

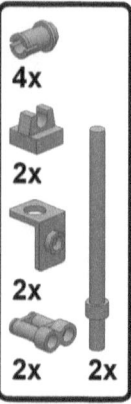
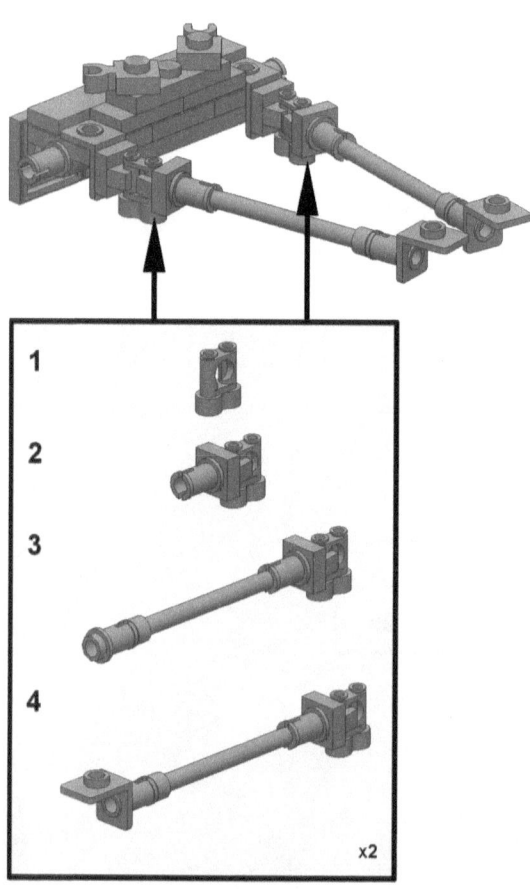

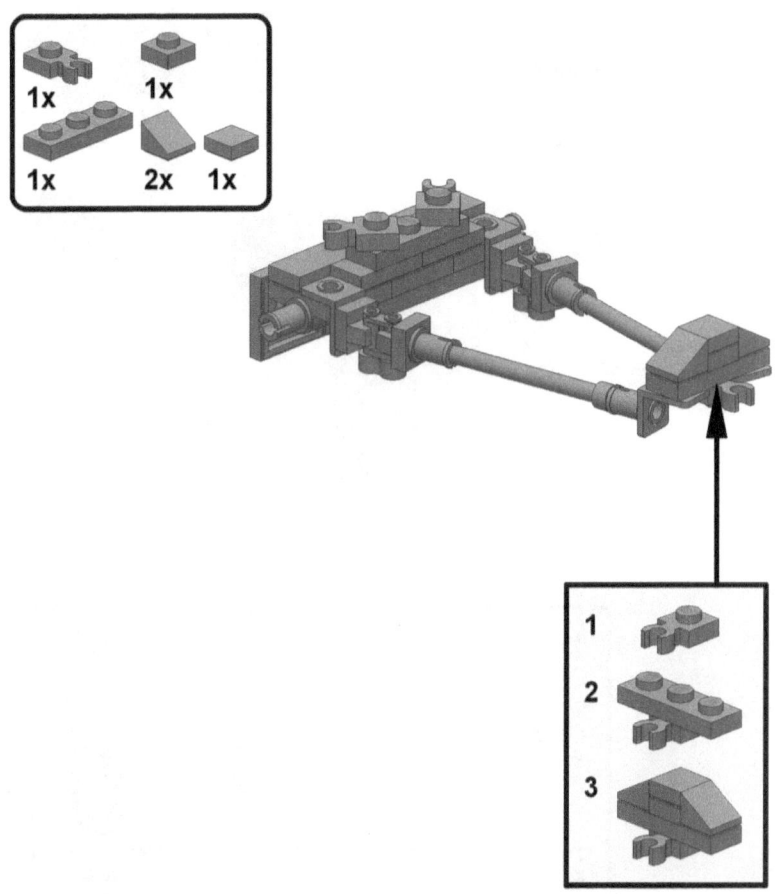

8

1x
1x
1x
1x
1x
1x
1x
1x
1x
2x

1
2
3
4
5
6
7

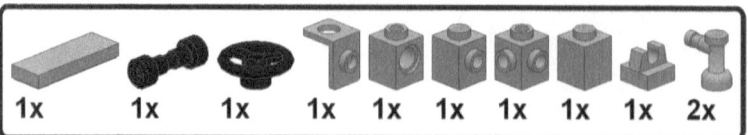

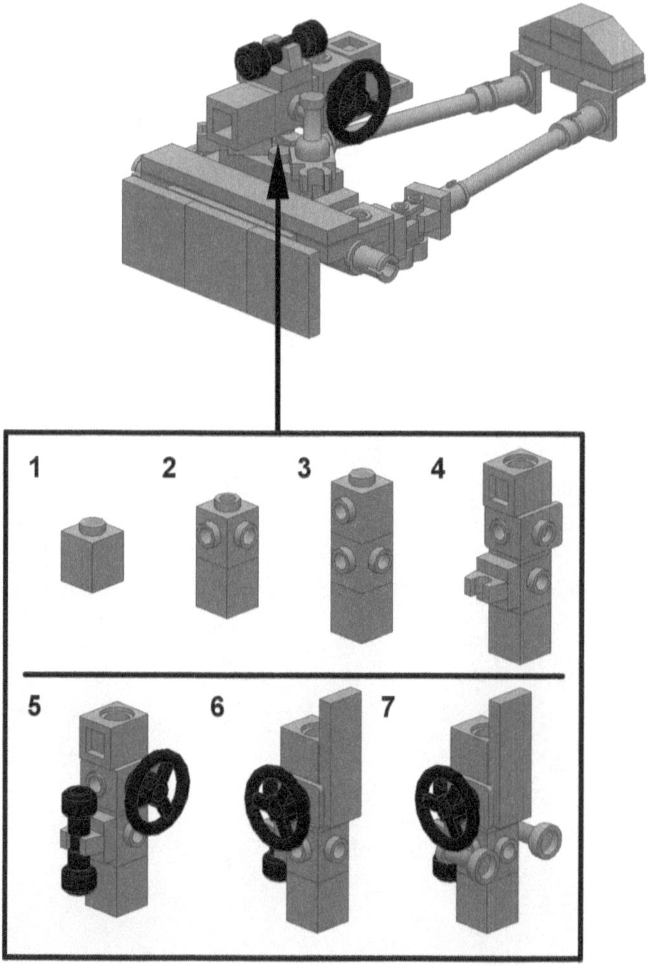

9

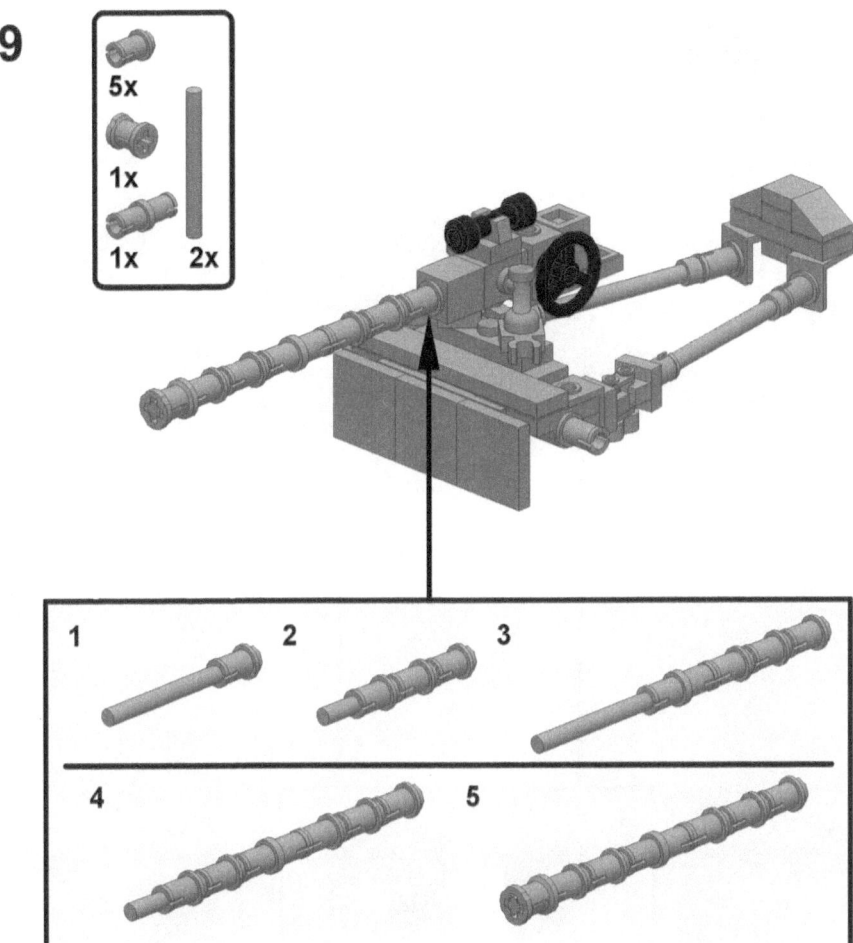

10

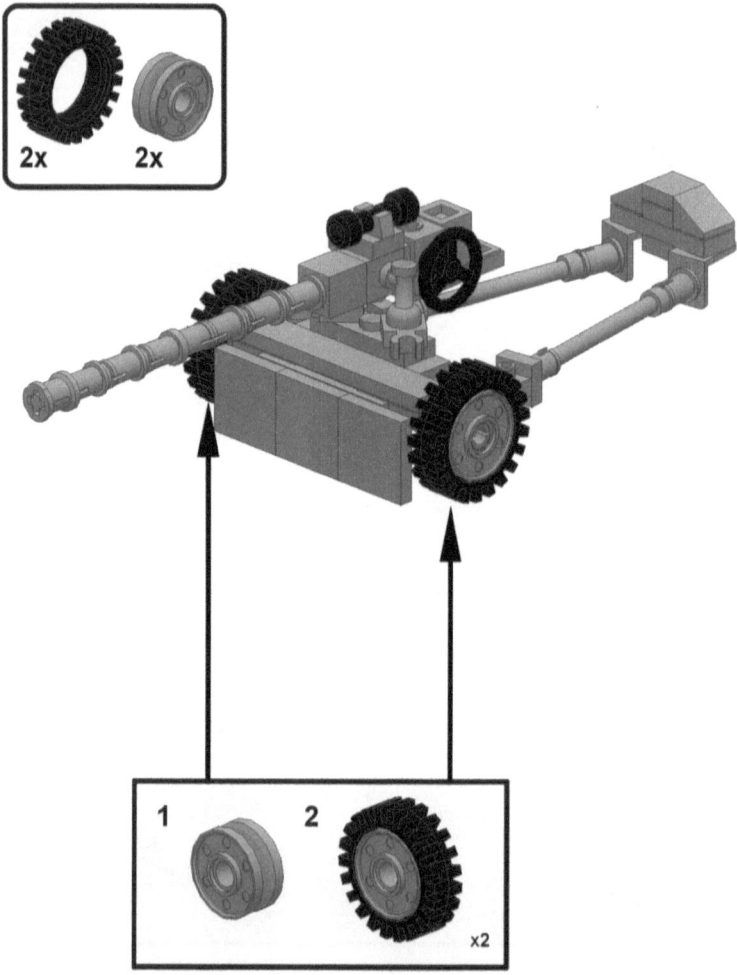

11

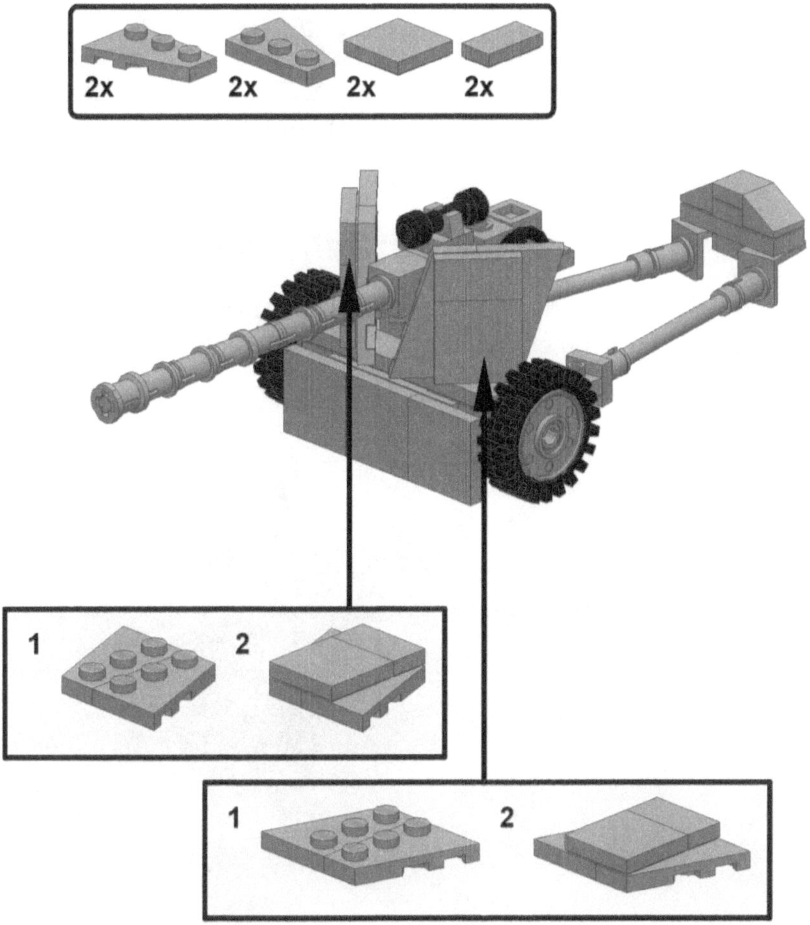

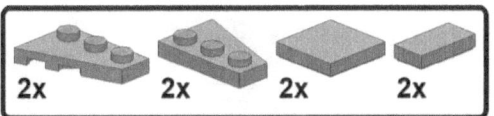

2cm Vierlings Flak

Zum Bau des Modells benötigen Sie ca. 73 LEGO® Bausteine.
Länge: ca. 10,0 cm Breite: ca. 7,0 cm Höhe: ca. 5,7 cm

2cm Anti-Aircraft Gun

Requires approx. 73 LEGO® bricks.
length: ca. 10.0 cm width: ca. 7.0 cm height: ca. 5.7 cm

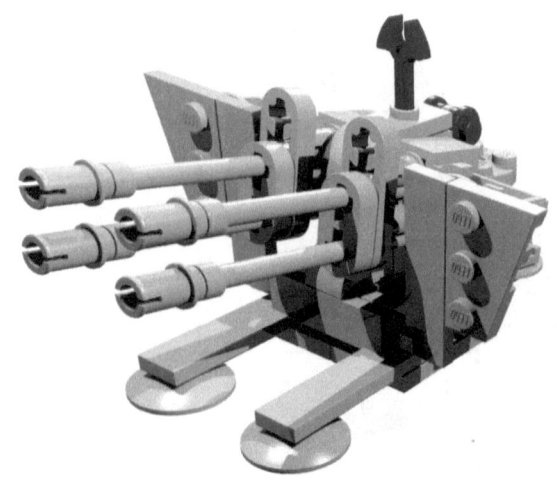

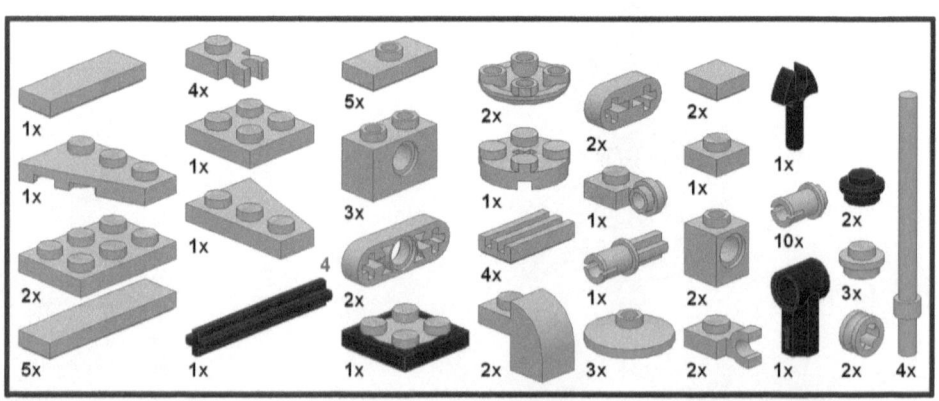

BLink ID	Color	Qty	Description
2412b	Grey	4	Tile, Modified 1 x 2 Grille with Bottom Groove
2431	Grey	5	Tile 1 x 4
4740	Grey	3	Dish 2 x 2 Inverted (Radar)
4073	Grey	3	Plate, Round 1 x 1 Straight Side
3022	Grey	1	Plate 2 x 2
3794	Grey	5	Plate, Modified 1 x 2 with 1 Stud (Jumper)
3680c02	Black	1	Turntable 2 x 2 Plate, Complete Assembly
3021	Grey	2	Plate 2 x 3
3070b	Grey	2	Tile 1 x 1 with Groove
43722	Grey	1	Wedge, Plate 3 x 2 Right
2654	Grey	2	Plate, Round 2 x 2 with Rounded Bottom
4265c	Grey	2	Technic Bush 1/2 Smooth
3705	Black	1	Technic, Axle 4
4073	Black	2	Plate, Round 1 x 1 Straight Side
4081b	Grey	1	Plate, Modified 1 x 1 with Clip Light
3700	Grey	3	Technic, Brick 1 x 2 with Hole
6541	Grey	2	Technic, Brick 1 x 1 with Hole
3749	Grey	1	Technic, Axle Pin without Friction Ridges Lengthwise
32013	Black	1	Technic, Axle and Pin Connector Angled #1
48729b	Black	1	Bar 1L with Clip Mechanical Claw, Cut Edges and Hole on One Side
6091	Grey	2	Brick, Modified 1 x 2 x 1 1/3 with Curved Top
3024	Grey	1	Plate 1 x 1
4274	Grey	10	Technic, Pin 1/2
6019	Grey	2	Plate, Modified 1 x 1 with Clip Horizontal
63864	Grey	1	Tile 1 x 3
4032b	Grey	1	Plate, Round 2 x 2 with Axle Hole
6632	Grey	2	Technic, Liftarm 1 x 3 Thin
63965	Grey	4	Bar 6L with Stop Ring
41677	Grey	2	Technic, Liftarm 1 x 2 Thin
4085	Grey	4	Plate, Modified 1 x 1 with Clip Vertical
43723	Grey	1	Wedge, Plate 3 x 2 Left

www.ww2custombrickmodels.de

1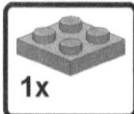

2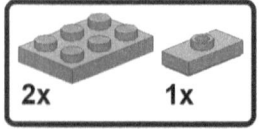

3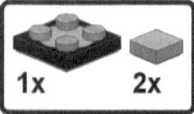

4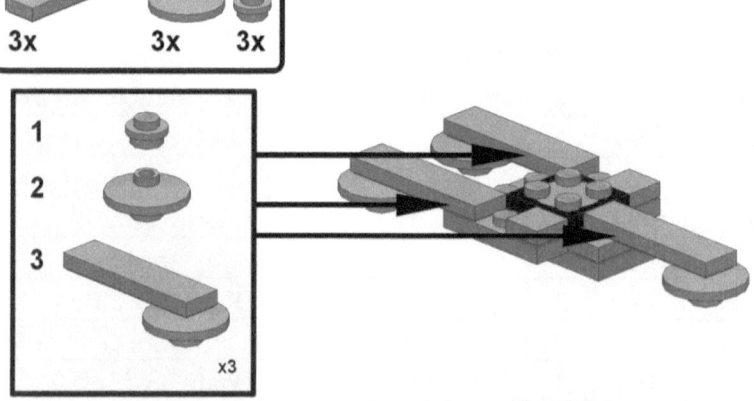

www.ww2custombrickmodels.de

5

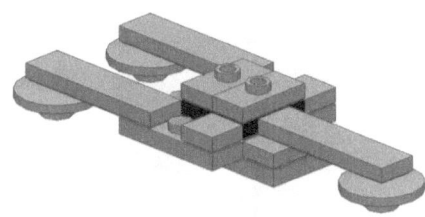

6

7

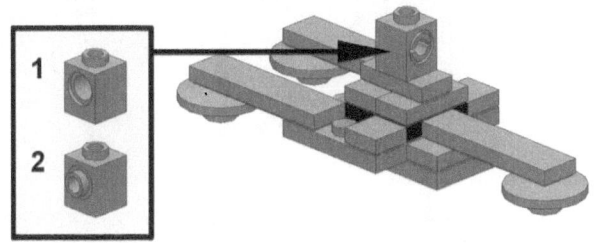

8

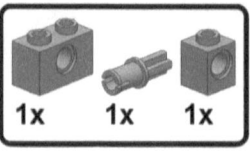

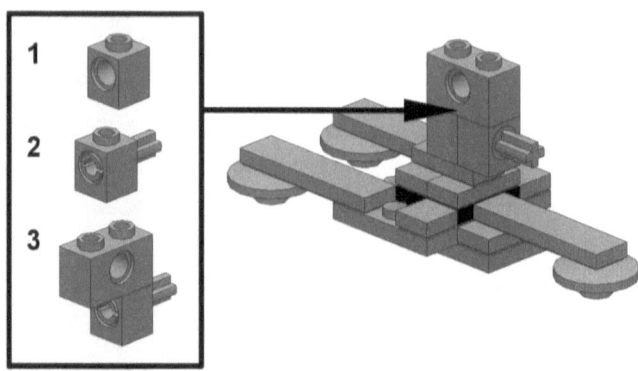

9

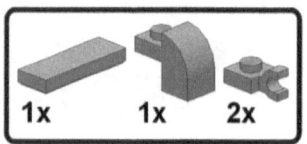

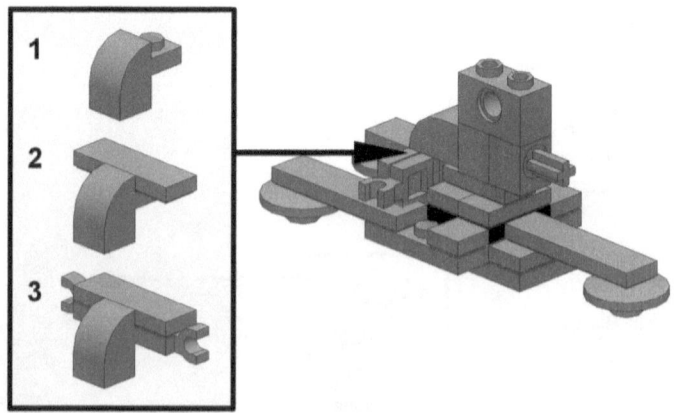

10

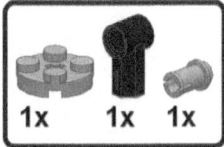

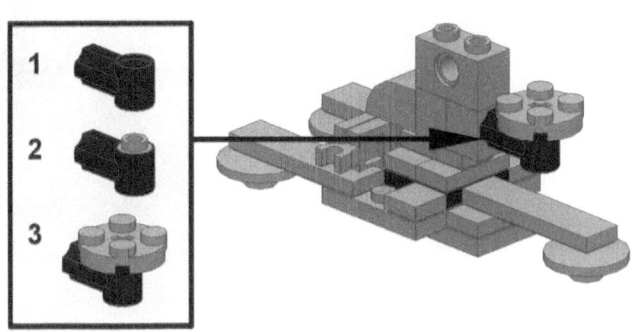

11

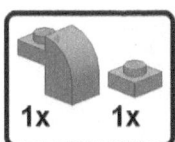

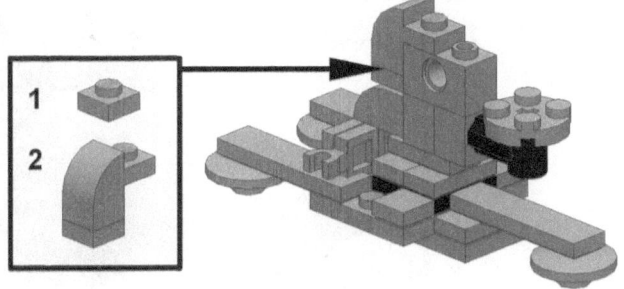

12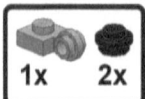

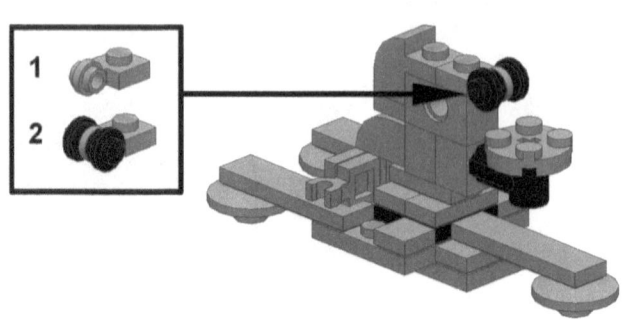

13

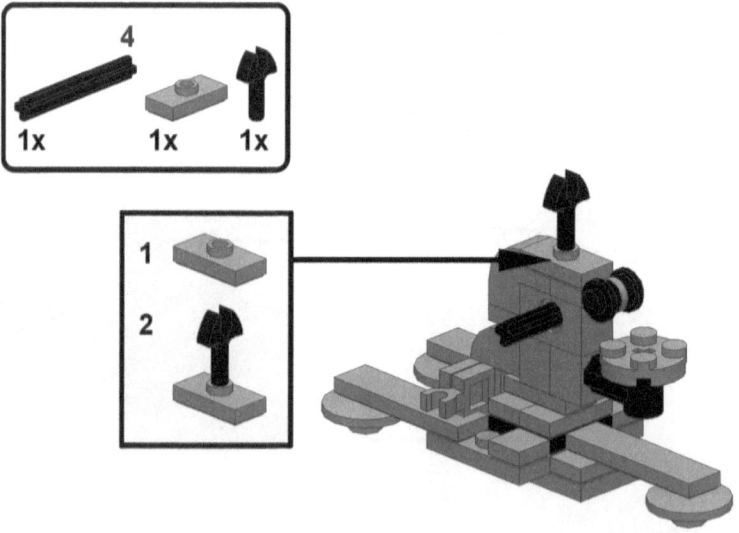

14

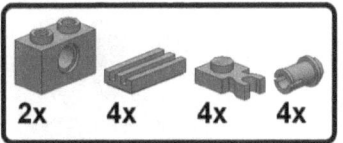
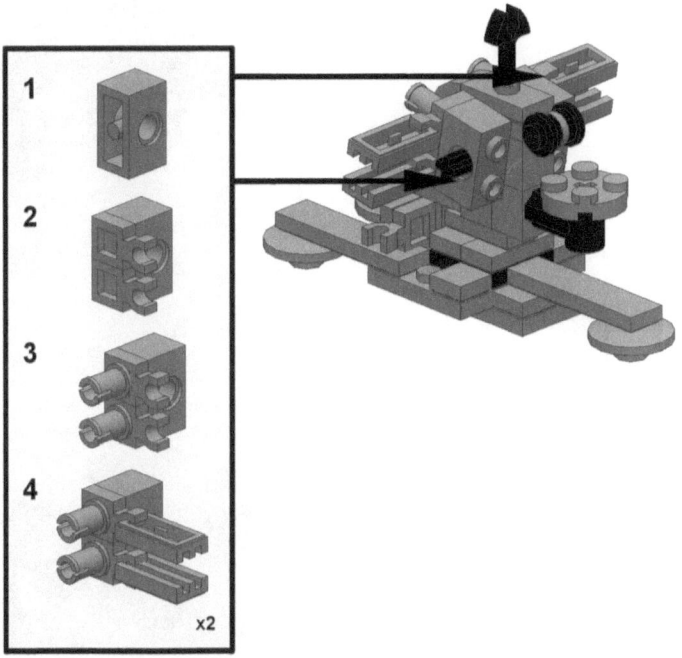

15

16

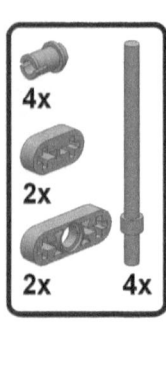
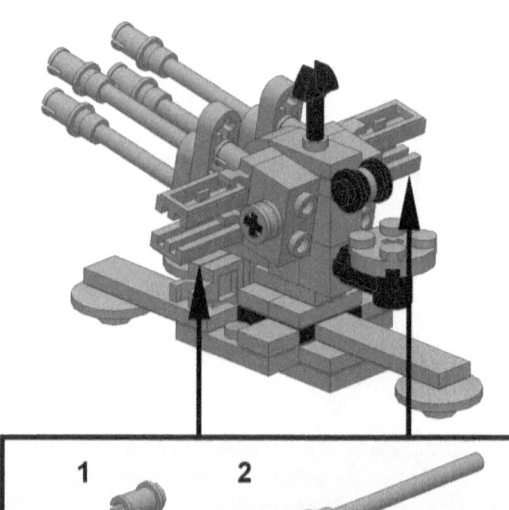
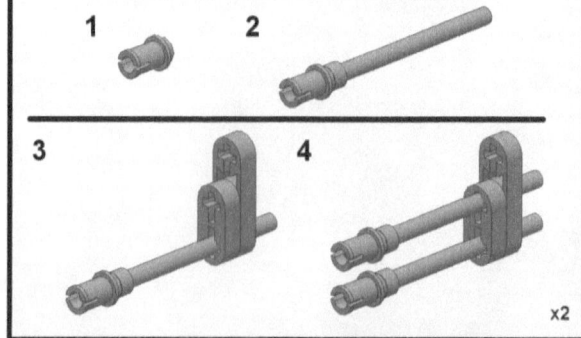

17

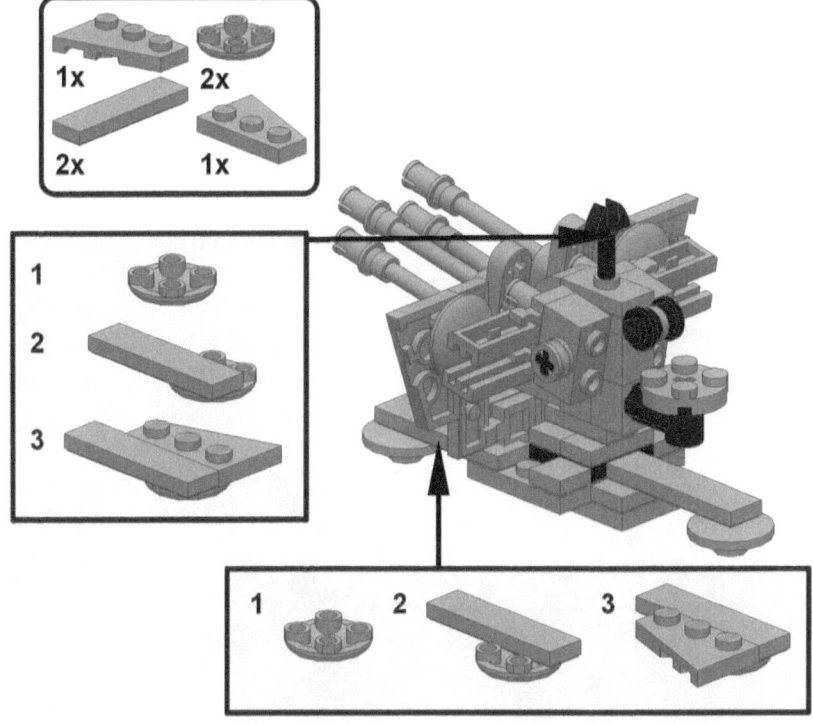

Motorrad mit Beiwagen

Zum Bau des Modells benötigen Sie ca. 95 LEGO® Bausteine.
Länge: ca. 9,4 cm Breite: ca. 7,2 cm Höhe: ca. 4,8 cm

Motorcycle with Sidecar

Requires approx. 95 LEGO® bricks.
length: ca. 9.4 cm width: ca. 7.2 cm height: ca. 4.8 cm

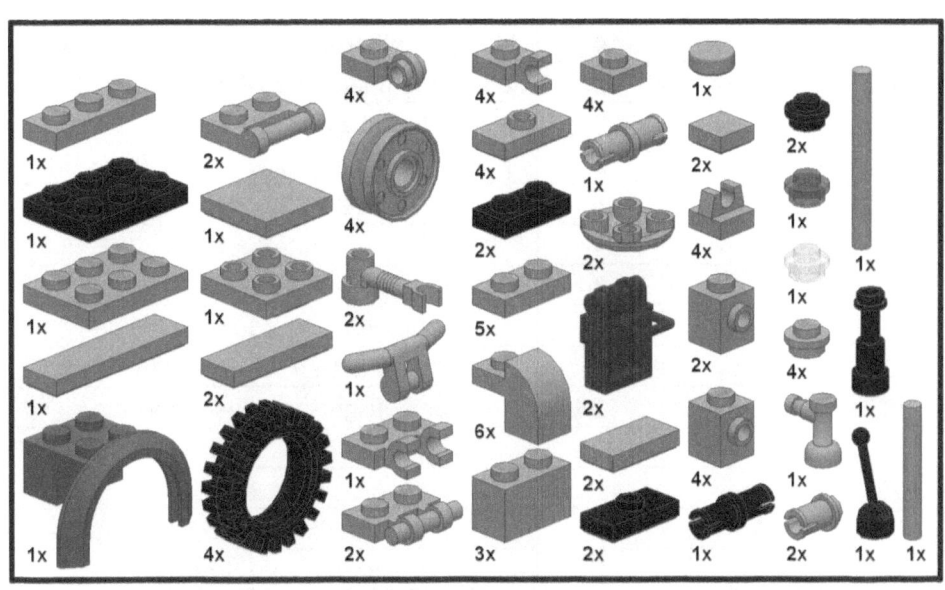

BLink ID	Color	Qty	Description
61254	Black	4	Tire Offset Tread
56902	Grey	4	Wheel 18mm D. x 8mm
64644	Black	1	Minifig, Utensil Telescope
4599b	Grey	1	Tap 1 x 1 without Hole in End
4073	Grey	4	Plate, Round 1 x 1 Straight Side
47905	Grey	2	Brick, Modified 1 x 1 with Studs on 2 Sides
3023	Black	2	Plate 1 x 2
3023	Grey	5	Plate 1 x 2
4081b	Grey	4	Plate, Modified 1 x 1 with Clip
4592c02	Grey	1	Lever Small Base with Black Lever
2555	Grey	4	Tile, Modified 1 x 1 with Clip
98138	Grey	1	Tile, Round 1 x 1
3794	Black	2	Plate, Modified 1 x 2 with 1 Stud (Jumper)
4073	Black	2	Plate, Round 1 x 1 Straight Side
3021	Black	1	Plate 2 x 3

3794	Grey	4	Plate, Modified 1 x 2 with 1 Stud (Jumper)
48336	Grey	2	Plate, Modified 1 x 2 with Handle on Side
2654	Grey	2	Plate, Round 2 x 2 with Rounded Bottom
2540	Grey	2	Plate, Modified 1 x 2 with Handle on Side
60470	Grey	1	Plate, Modified 1 x 2 with Clips Horizontal
3024	Grey	4	Plate 1 x 1
2524	Brown	2	Minifig, Backpack Non-Opening
4073	Tr. Red	1	Plate, Round 1 x 1 Straight Side
6019	Grey	4	Plate, Modified 1 x 1 with Clip Horizontal
30374	Grey	1	Bar 4L (Lightsaber Blade / Wand)
4274	Grey	2	Technic, Pin 1/2
87994	Grey	1	Bar 3L
4735	Grey	2	Bar 1 x 3 with Clip and Stud Receptacle
4073	Transparent	1	Plate, Round 1 x 1 Straight Side
30031	Grey	1	Minifig, Utensil Handlebars
63864	Grey	2	Tile 1 x 3
87087	Grey	4	Brick, Modified 1 x 1 with Stud on 1 Side
3070b	Grey	2	Tile 1 x 1 with Groove
2780	Black	1	Technic, Pin with Friction
3673	Grey	1	Technic, Pin without Friction
2444	Grey	1	Plate, Modified 2 x 2 with Pin Hole
3021	Grey	1	Plate 2 x 3
3623	Grey	1	Plate 1 x 3
3068b	Grey	1	Tile 2 x 2 with Groove
50745	Dark Grey	1	Vehicle, Mudguard 4 x 2 1/2 x 2
3004	Grey	3	Brick 1 x 2
2431	Grey	1	Tile 1 x 4
6091	Grey	6	Brick, Modified 1 x 2 x 1 1/3 with Curved Top
3069b	Grey	2	Tile 1 x 2 with Groove

1

2

3

4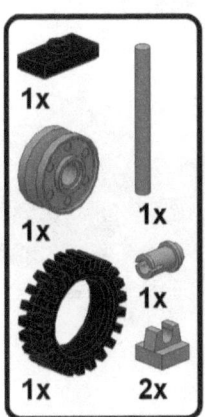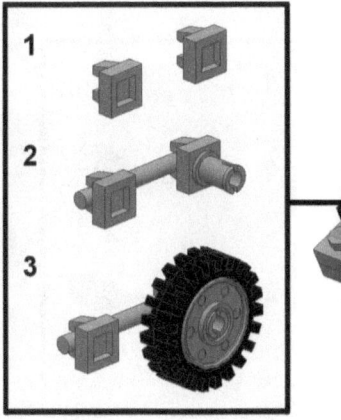

www.ww2custombrickmodels.de

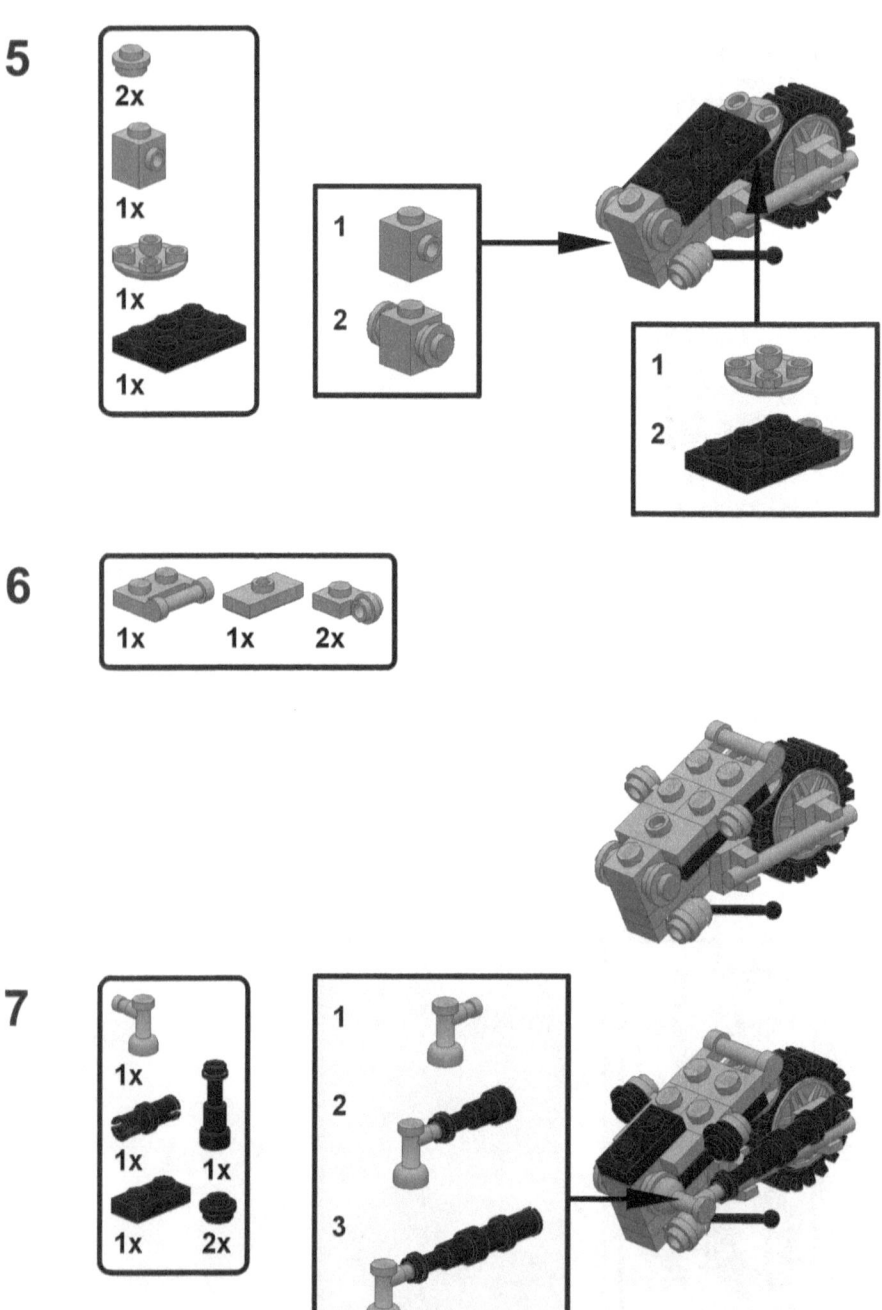

8

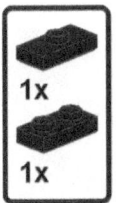

9

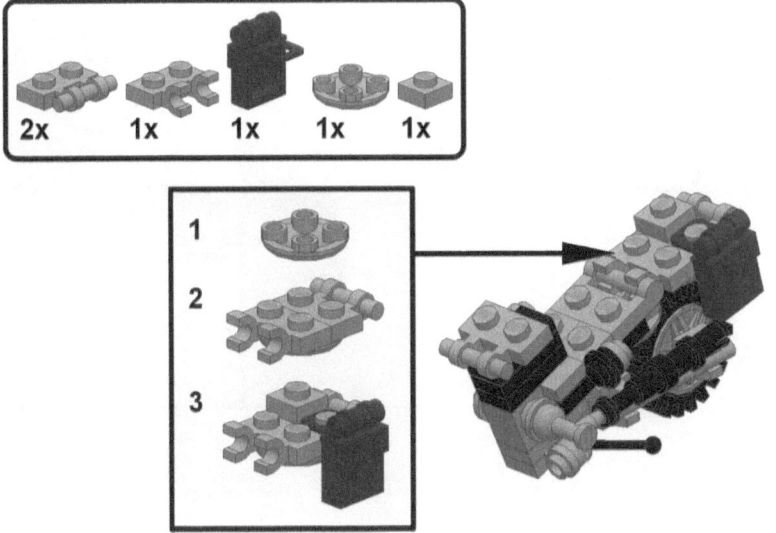

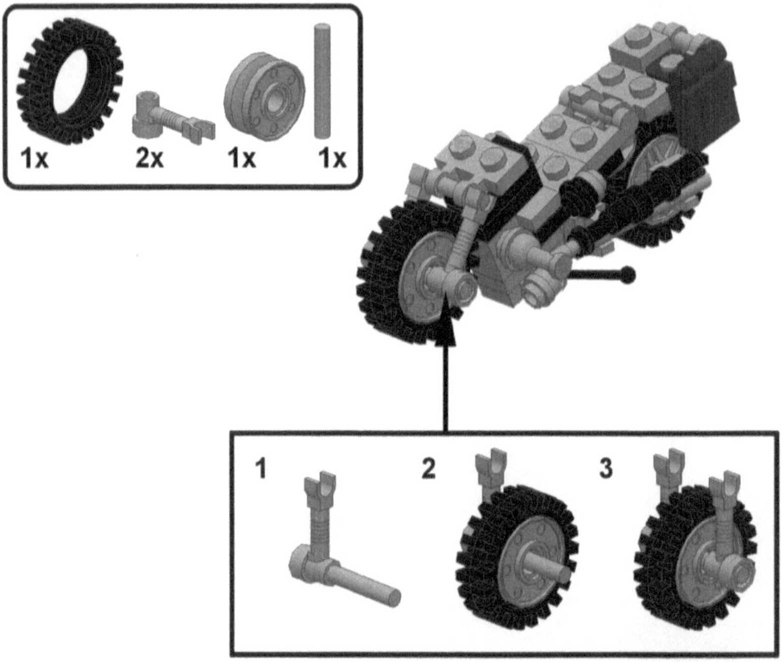

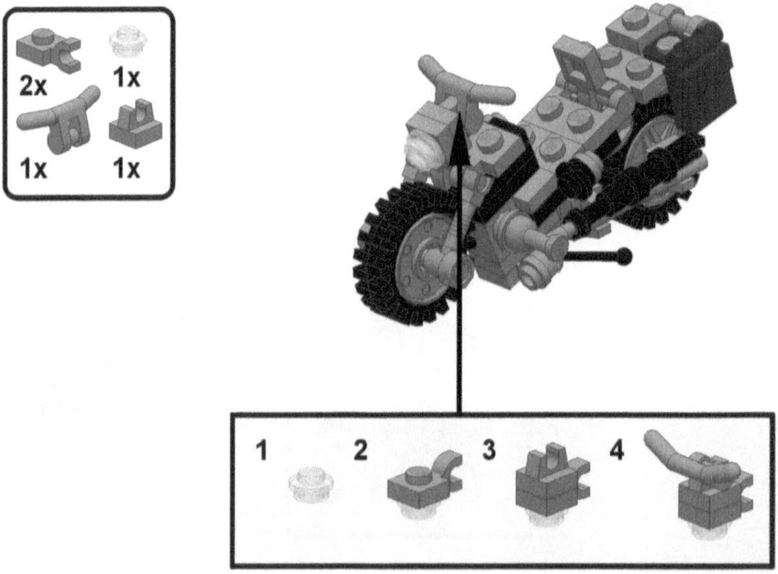

12

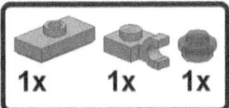

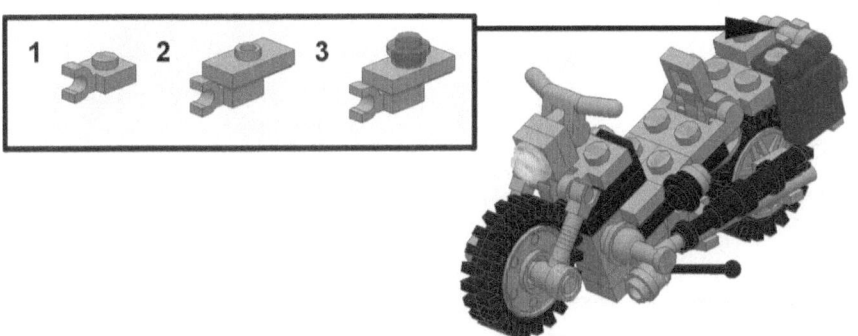

1

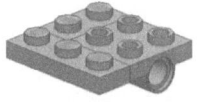

2

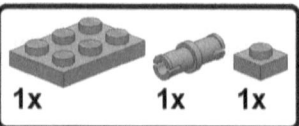

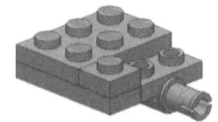

3

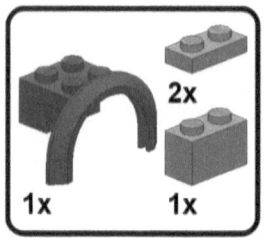

4

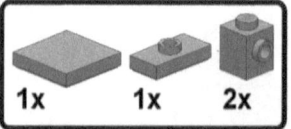

www.ww2custombrickmodels.de

5

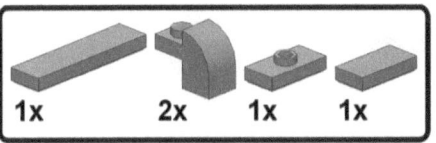

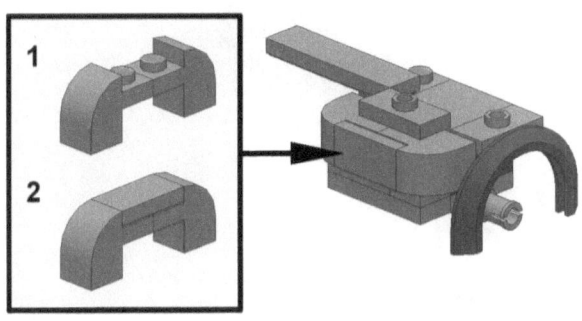

6

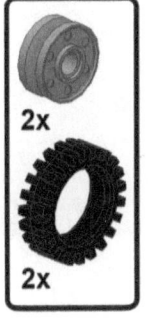

www.ww2custombrickmodels.de

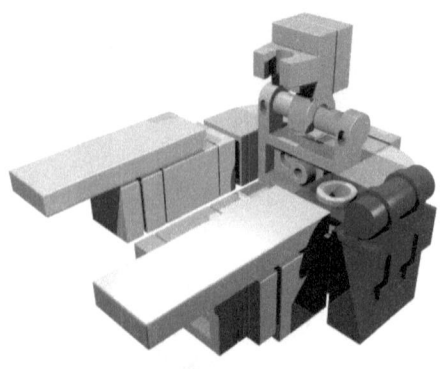

1

2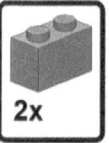

3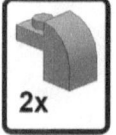

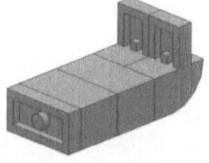

www.ww2custombrickmodels.de

4

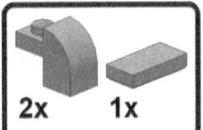

5

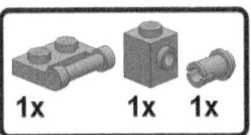

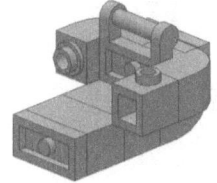

6

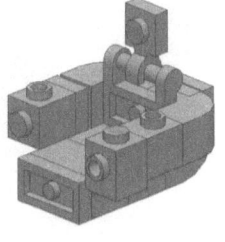

www.ww2custombrickmodels.de

7

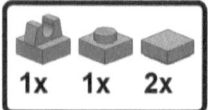

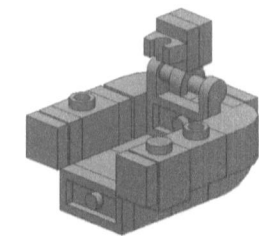

8

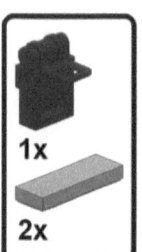

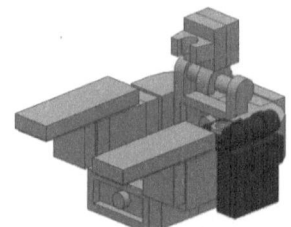

7

13

Kradmelder Motorrad

Zum Bau des Modells benötigen Sie ca. 95 LEGO® Bausteine.
Länge: ca. 9,6 cm Breite: ca. 3,7 cm Höhe: ca. 4,3 cm

Single Seat Motorcycle

Requires approx. 95 LEGO® bricks.
length: ca. 9.6 cm width: ca. 3.7 cm height: ca. 4.3 cm

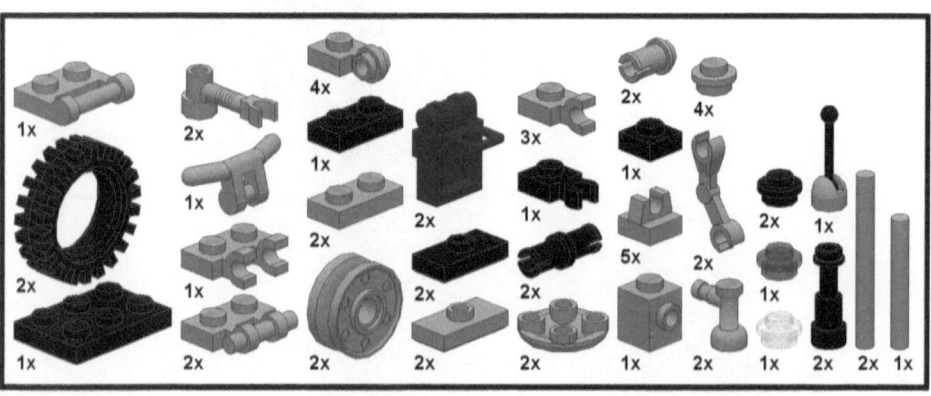

BLink ID	Color	Qty	Description
4081b	Grey	4	Plate, Modified 1 x 1 with Clip
4073	Grey	4	Plate, Round 1 x 1 Straight Side
2555	Grey	5	Tile, Modified 1 x 1 with Clip
3023	Grey	2	Plate 1 x 2
4592c02	Grey	1	Lever Small Base with Black Lever
47905	Grey	1	Brick, Modified 1 x 1 with Studs on 2 Sides
4599b	Grey	2	Tap 1 x 1 without Hole in End
64644	Black	2	Minifig, Utensil Telescope
3023	Black	1	Plate 1 x 2
3024	Black	1	Plate 1 x 1
3794	Black	2	Plate, Modified 1 x 2 with 1 Stud (Jumper)
4085c	Black	1	Plate, Modified 1 x 1 with Clip Vertical
30374	Grey	2	Bar 4L (Lightsaber Blade / Wand)
3021	Black	1	Plate 2 x 3
3794	Grey	2	Plate, Modified 1 x 2 with 1 Stud (Jumper)
4073	Black	2	Plate, Round 1 x 1 Straight Side
2654	Grey	2	Plate, Round 2 x 2 with Rounded Bottom
48336	Grey	1	Plate, Modified 1 x 2 with Handle on Side
6019	Grey	3	Plate, Modified 1 x 1 with Clip Horizontal
2540	Grey	2	Plate, Modified 1 x 2 with Handle on Side
60470	Grey	1	Plate, Modified 1 x 2 with Clips Horizontal
2524	Brown	2	Minifig, Backpack Non-Opening
4073	Tr. Red	1	Plate, Round 1 x 1 Straight Side
30377	Grey	2	Arm Mechanical, Battle Droid
4735	Grey	2	Bar 1 x 3 with Clip and Stud Receptacle
87994	Grey	1	Bar 3L
4274	Grey	2	Technic, Pin 1/2
61254	Black	2	Tire Offset Tread - Band Around Center of Tread
56902	Grey	2	Wheel 18mm D. x 8mm
4073	Transparent	1	Plate, Round 1 x 1 Straight Side
30031	Grey	1	Minifig, Utensil Handlebars
2780	Black	2	Technic, Pin with Friction Ridges Lengthwise

www.ww2custombrickmodels.de

1

2

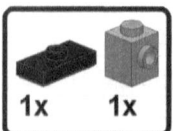

3

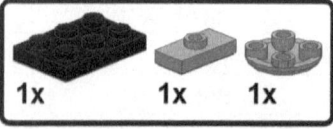

www.ww2custombrickmodels.de

4

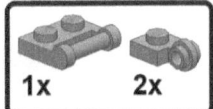

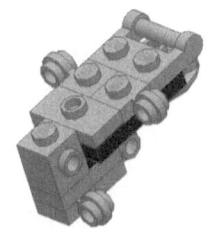

5

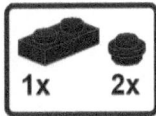

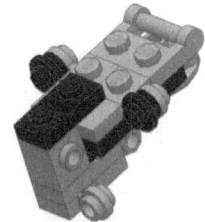

6

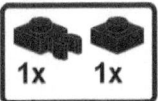

7

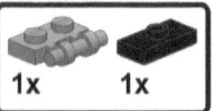

8

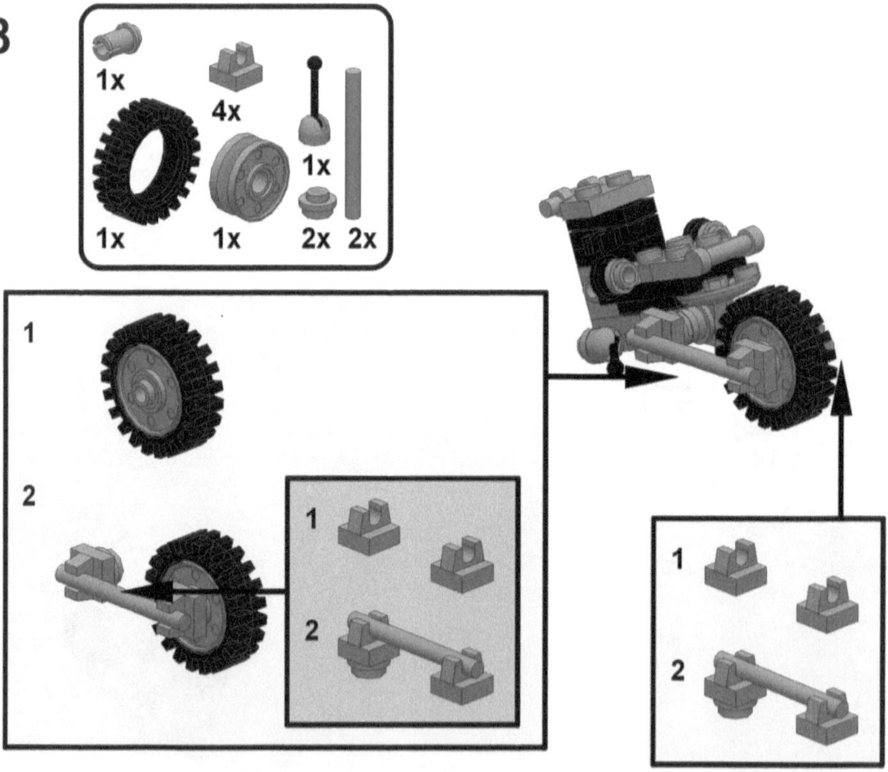

9

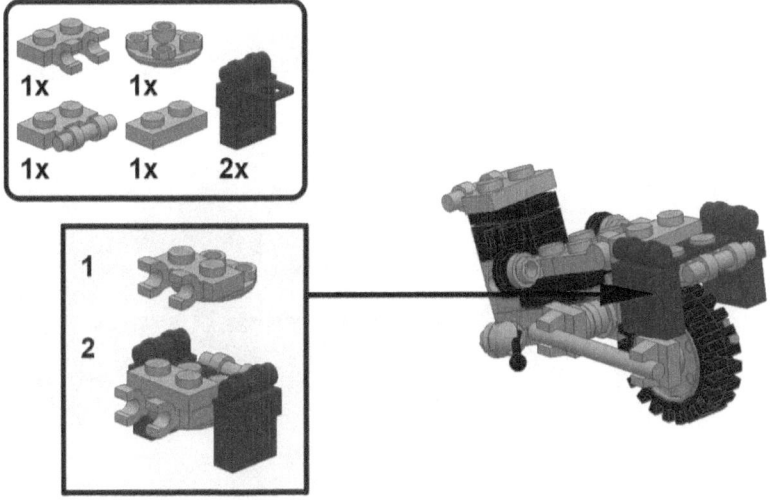

10

11

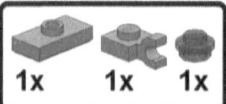

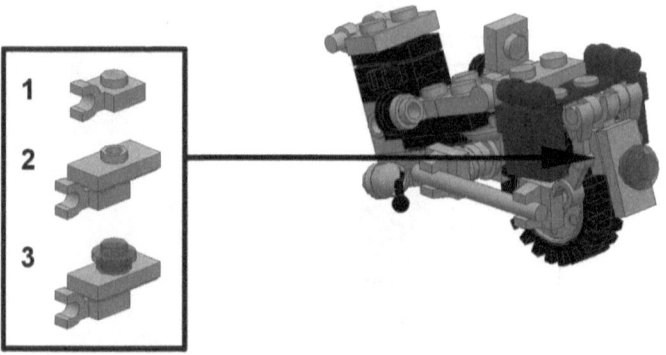

12

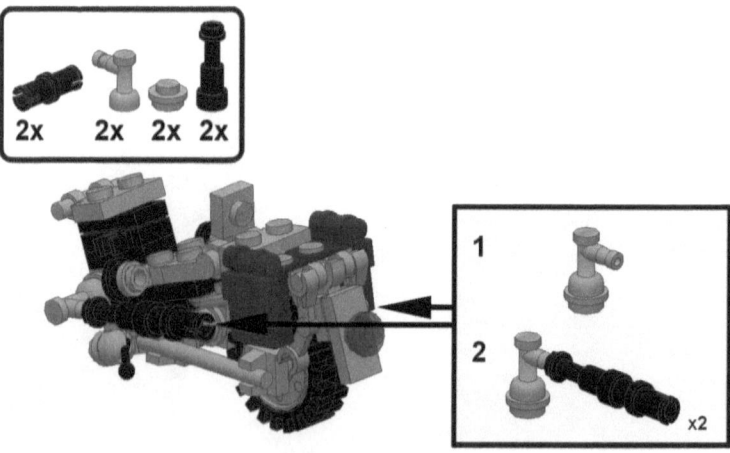

13

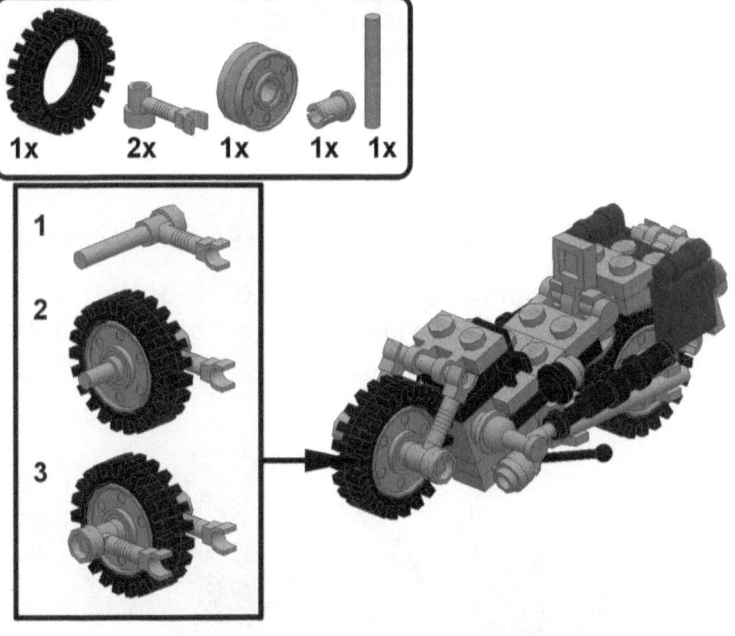

14

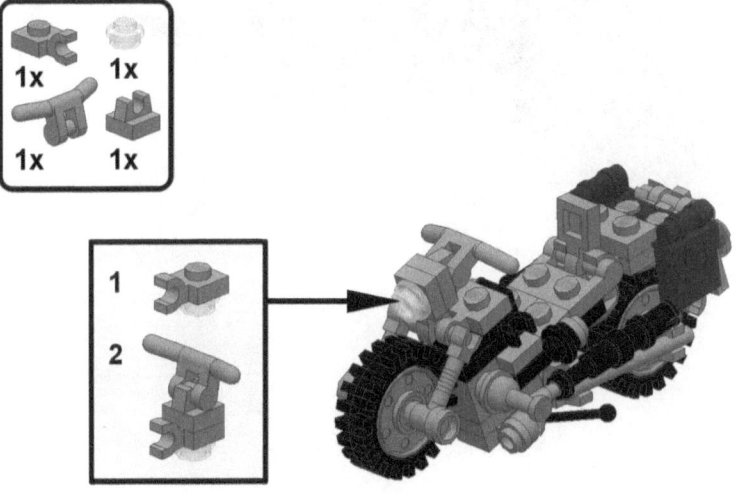

NSU Kettenkrad

Zum Bau des Modells benötigen Sie ca. 212 LEGO® Bausteine.
Länge: ca. 20,4 cm Breite: ca. 4,8 cm Höhe: ca. 4,9 cm

Chain Motorcycle

Requires approx. 212 LEGO® bricks.
length: ca. 20.4 cm width: ca. 4.8 cm height: ca. 4.9 cm

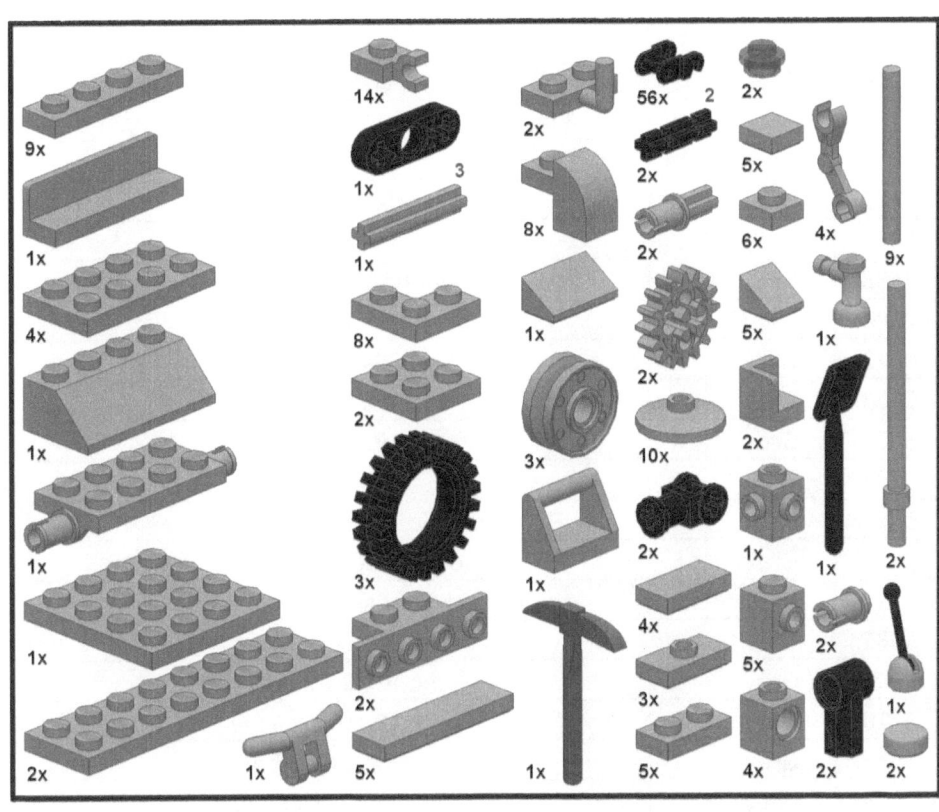

BLink ID	Color	Qty	Description
4274	Grey	2	Technic, Pin 1/2
3794	Grey	3	Plate, Modified 1 x 2 with 1 Stud (Jumper)
6019	Grey	14	Plate, Modified 1 x 1 with Clip Horizontal
2420	Grey	8	Plate 2 x 2 Corner
6231	Grey	2	Panel 1 x 1 x 1 Corner
2436	Grey	2	Bracket 1 x 2 - 1 x 4
4599b	Grey	1	Tap 1 x 1 without Hole in End
3022	Grey	2	Plate 2 x 2
3023	Grey	5	Plate 1 x 2
3020	Grey	4	Plate 2 x 4
3034	Grey	2	Plate 2 x 8
6541	Grey	4	Technic, Brick 1 x 1 with Hole
47905	Grey	5	Brick, Modified 1 x 1 with Studs on 2 Sides
4623	Grey	2	Plate, Modified 1 x 2 with Arm Up
3031	Grey	1	Plate 4 x 4
85984	Grey	1	Slope 30 1 x 2 x 2/3
3710	Grey	9	Plate 1 x 4
3037	Grey	1	Slope 45 2 x 4
3070b	Grey	5	Tile 1 x 1 with Groove
2431	Grey	5	Tile 1 x 4
3024	Grey	6	Plate 1 x 1
54200	Grey	5	Slope 30 1 x 1 x 2/3
3069b	Grey	4	Tile 1 x 2 with Groove
2432	Grey	1	Tile, Modified 1 x 2 with Handle
30413	Grey	1	Panel 1 x 4 x 1
30374	Grey	9	Bar 4L (Lightsaber Blade / Wand)
4740	Grey	10	Dish 2 x 2 Inverted (Radar)
4019	Grey	2	Technic, Gear 16 Tooth
3749	Grey	2	Technic, Axle Pin without Friction
30377	Grey	4	Arm Mechanical, Battle Droid
98138	Grey	2	Tile, Round 1 x 1
3837	Black	1	Minifig, Utensil Shovel (Round Stem End)

32013	Black	2	Technic, Axle and Pin Connector Angled #1
32062	Black	2	Technic, Axle 2 Notched
32039	Black	2	Technic, Axle Connector with Axle Hole
4519	Grey	1	Technic, Axle 3
3841	Dark Grey	1	Minifig, Utensil Pickaxe
4592c02	Grey	1	Lever Small Base with Black Lever
4733	Grey	1	Brick, Modified 1 x 1 with Studs on 4 Sides
30031	Grey	1	Minifig, Utensil Handlebars
30157	Grey	1	Plate, Modified 2 x 4 with Pins
6091	Grey	8	Brick, Modified 1 x 2 x 1 1/3 with Curved Top
4073	Tr. Red	2	Plate, Round 1 x 1 Straight Side
6632	Black	1	Technic, Liftarm 1 x 3 Thin
63965	Grey	2	Bar 6L with Stop Ring
3711	Black	56	Technic, Link Chain
61254	Black	3	Tire Offset Tread - Band Around Center of Tread
56902	Grey	3	Wheel 18mm D. x 8mm

1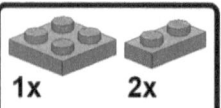

2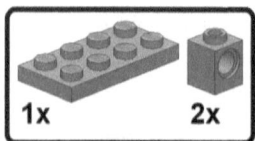

3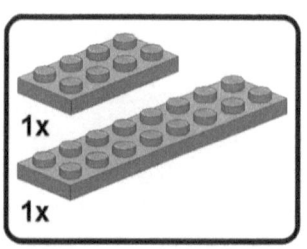

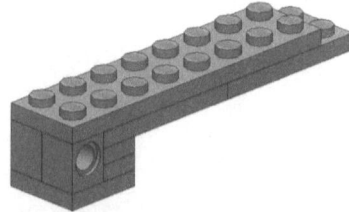

4

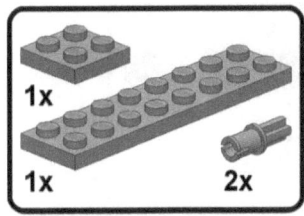

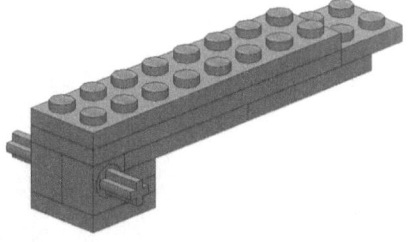

5

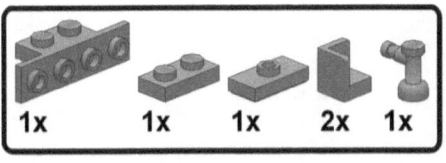

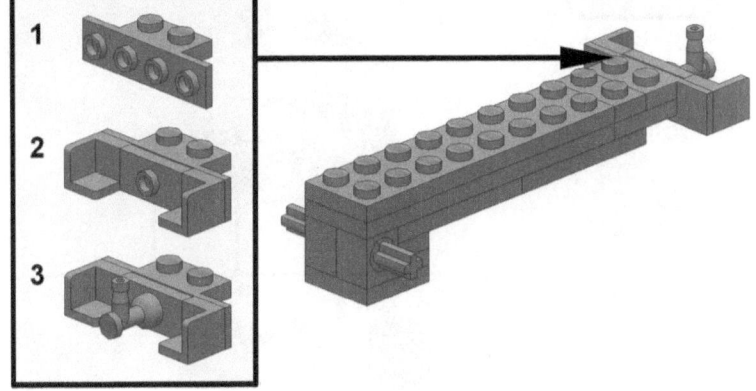

6

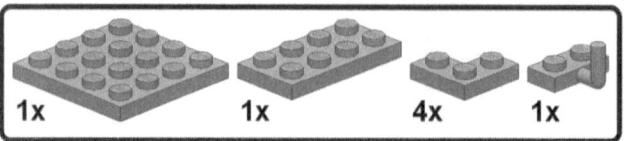
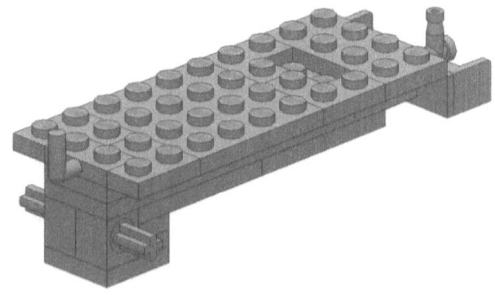

7

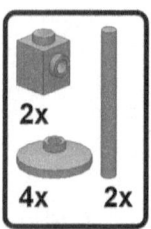
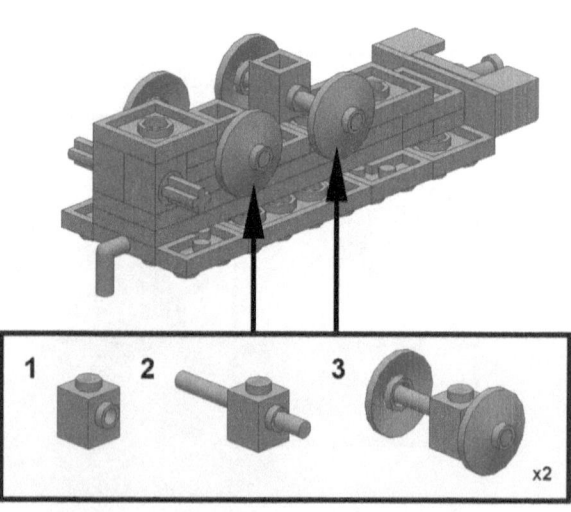

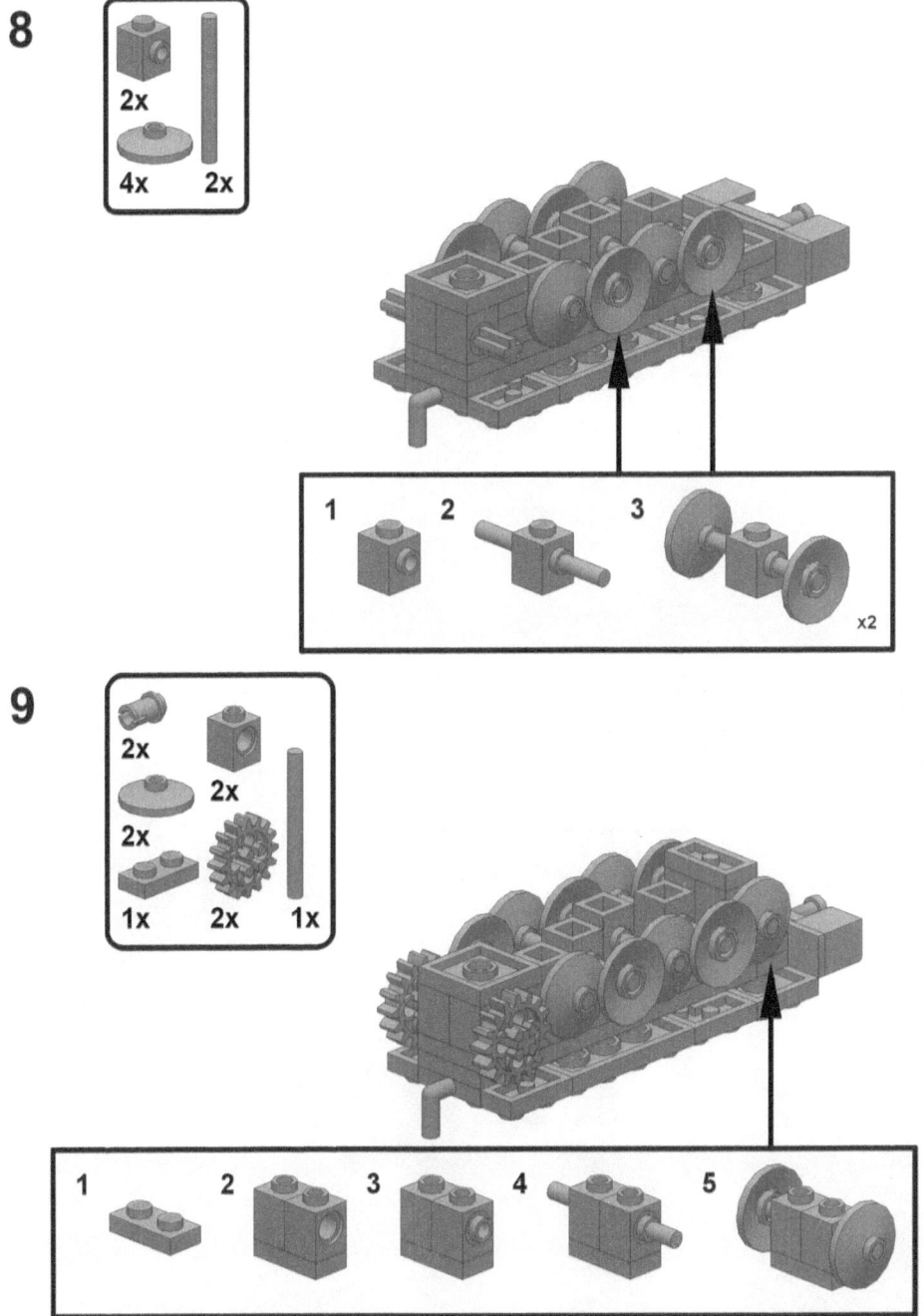

10

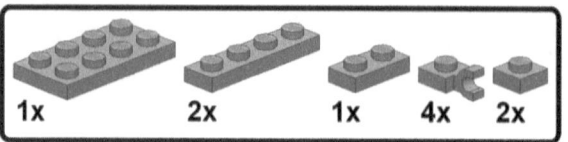

11

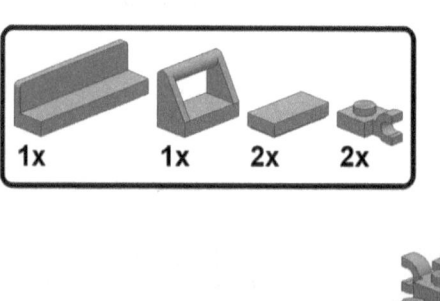

12

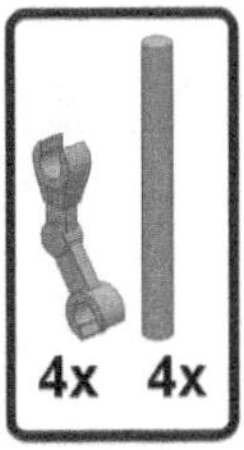

4x 4x

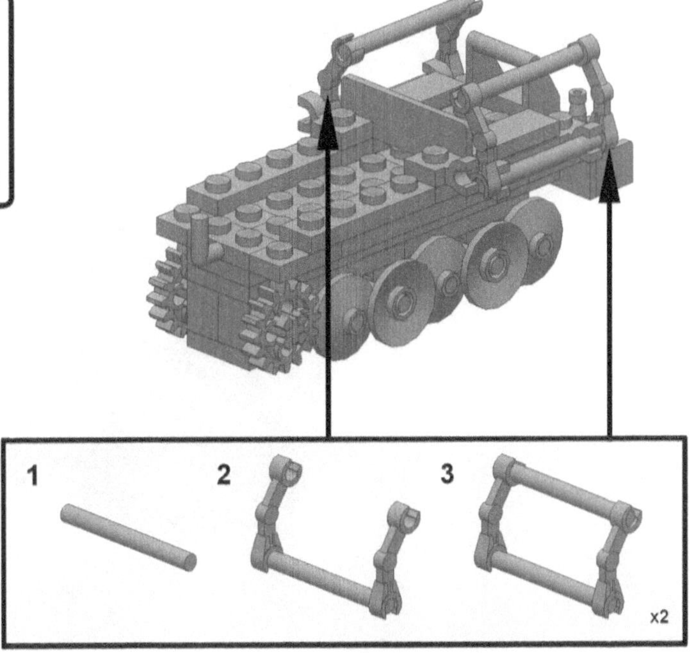

1 2 3

x2

13

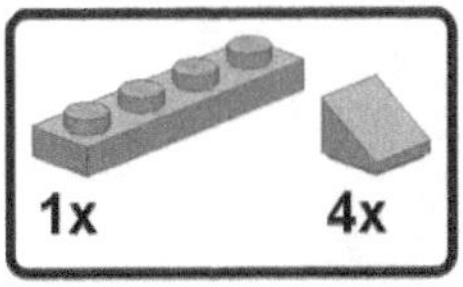

1x 4x

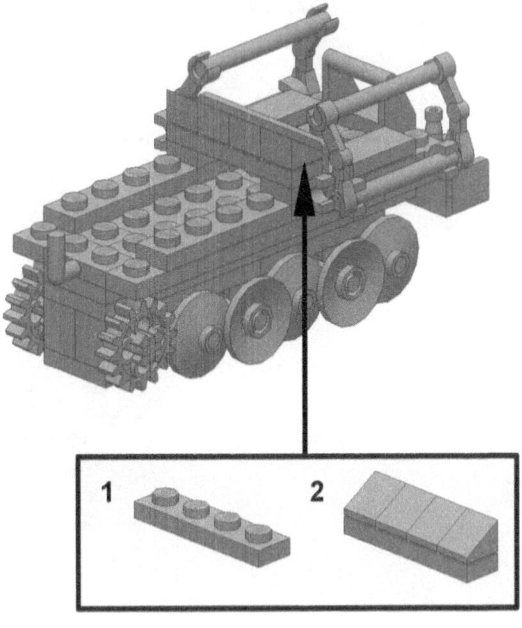

1 2

www.ww2custombrickmodels.de

59

14

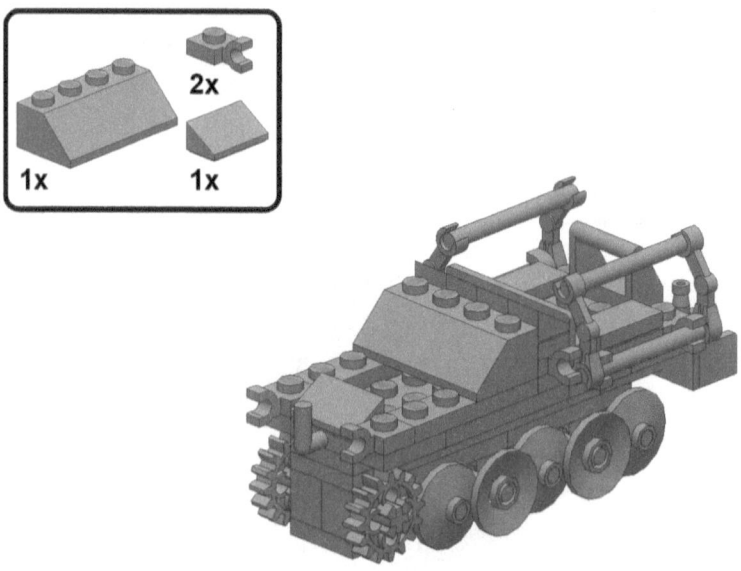

15

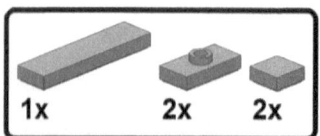

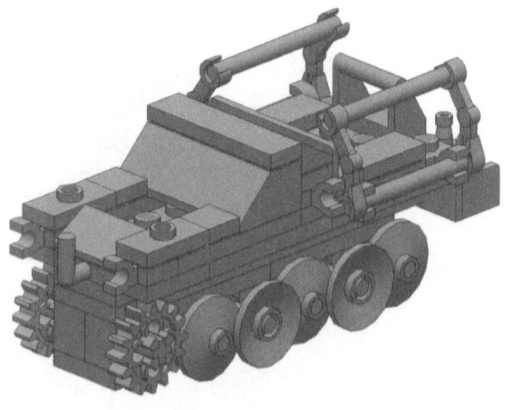

16

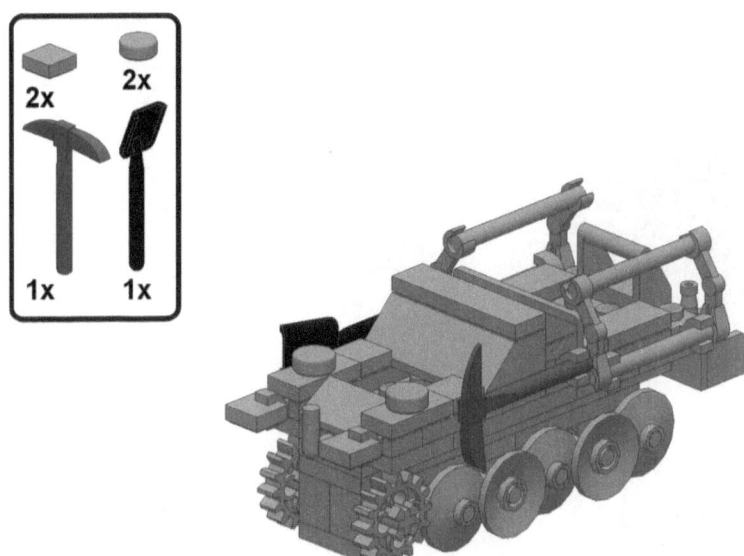

1

2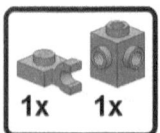

3

www.ww2custombrickmodels.de

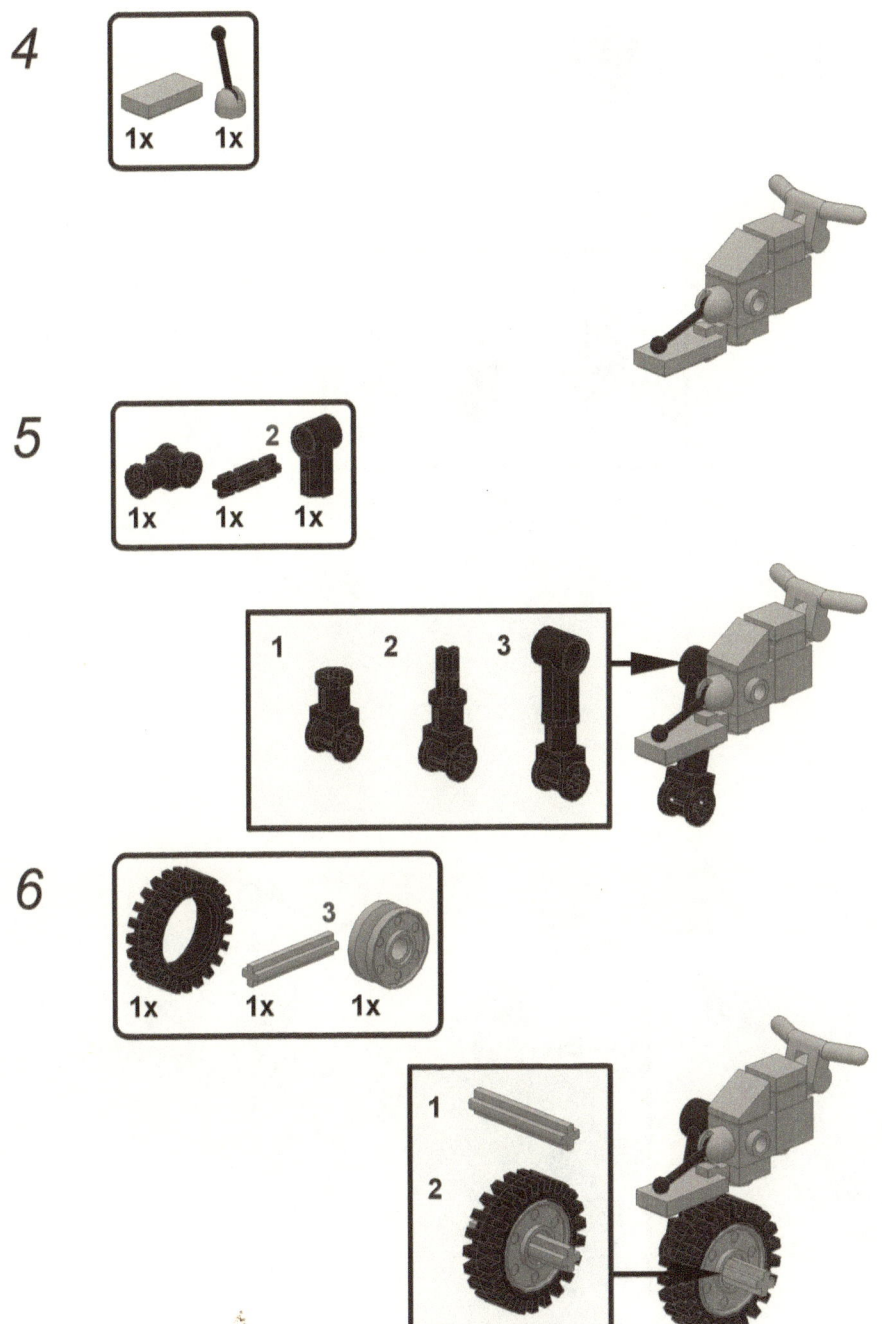

7

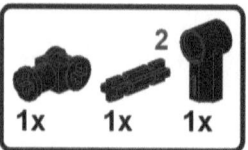

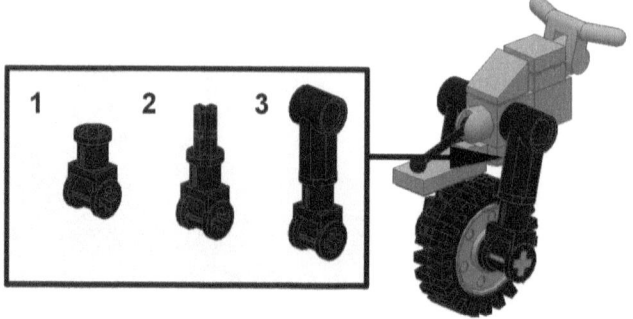

17

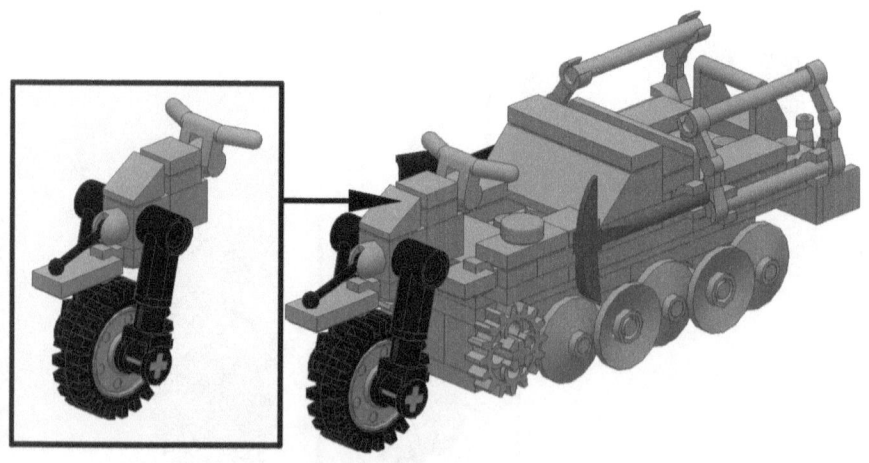

1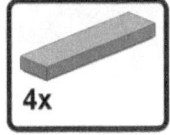

2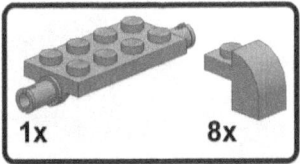

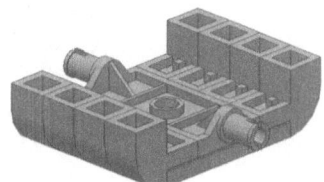

3

www.ww2custombrickmodels.de

4

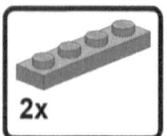

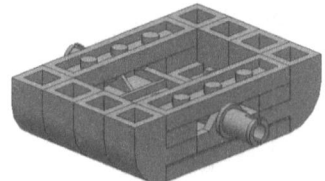

5

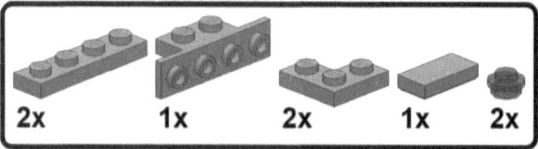

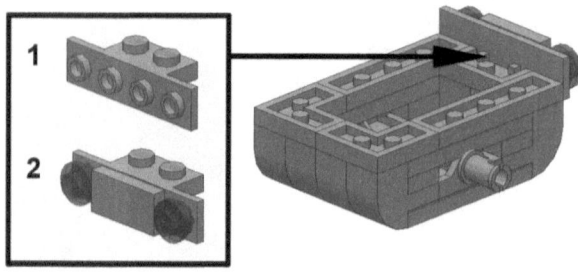

6

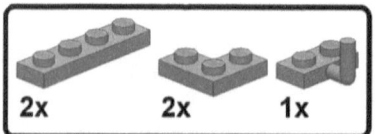

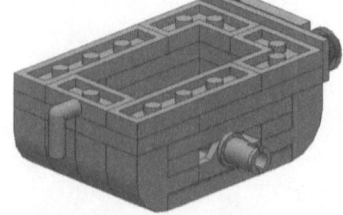

7

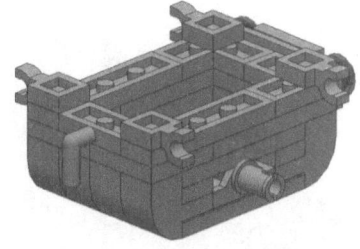

8

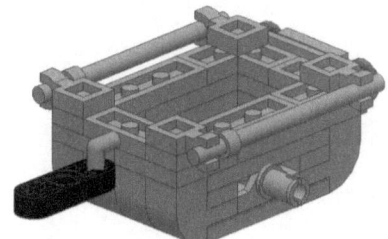

9

19

Sie sind auf der Suche nach den Teilen? Besuchen Sie einige der unten stehenden Internetseiten um mit der Hilfe der aufgelisteten Teilenummern dort das passende Teil zu finden.

Not sure where to find the parts? Try some of the websites below. You can search for the parts with the listed part numbers.

www.lego.de

www.bricklink.com

www.brickowl.com

www.ebay.de

![Logo: www.ww2custombrickmodels.de by Martin Ludwig]

Visit my website:

www.ww2custombrickmodels.de

www.ingramcontent.com/pod-product-compliance
Lightning Source LLC
Chambersburg PA
CBHW030456220526
45464CB00006B/2562